SPACE SCIENCE

PLANETS

BY BETSY RATHBURN

BELLWETHER MEDIA • MINNEAPOLIS, MN

Are you ready to take it to the extreme? Torque books thrust you into the action-packed world of sports, vehicles, mystery, and adventure. These books may include dirt, smoke, fire, and chilling tales. **WARNING**: read at your own risk.

This edition first published in 2019 by Bellwether Media, Inc.

No part of this publication may be reproduced in whole or in part without written permission of the publisher.
For information regarding permission, write to Bellwether Media, Inc.,
Attention: Permissions Department,
6012 Blue Circle Drive, Minnetonka, MN 55343.

Library of Congress Cataloging-in-Publication Data

Names: Rathburn, Betsy, author.
Title: Planets / by Betsy Rathburn.
Description: Minneapolis, MN : Bellwether Media, Inc., [2019] | Series:
   Torque: Space Science | Audience: Ages 7-12. | Audience: Grades 3 to 7. |
   Includes bibliographical references and index.
Identifiers: LCCN 2018001133 (print) | LCCN 2018007960 (ebook) | ISBN
   9781681036021 (ebook) | ISBN 9781626178618 (hardcover : alk. paper)
Subjects: LCSH: Planets–Juvenile literature. | Planetary science–Juvenile
   literature.
Classification: LCC QB602 (ebook) | LCC QB602 .R377 2019 (print) | DDC
   523.4–dc23
LC record available at https://lccn.loc.gov/2018001133

Editor: Rebecca Sabelko   Designer: Andrea Schneider

Printed in the United States of America, North Mankato, MN.

TABLE OF CONTENTS

LOOKING FOR LIFE

It is April 17, 2014. **Astronomers** have just announced the discovery of Kepler-186f. This is the first Earth-sized planet found in another star's **habitable zone**.

Scientists conduct many studies of this **exoplanet**. What is the planet made of? How hot is it? The answers to these questions may one day help scientists find life on other worlds!

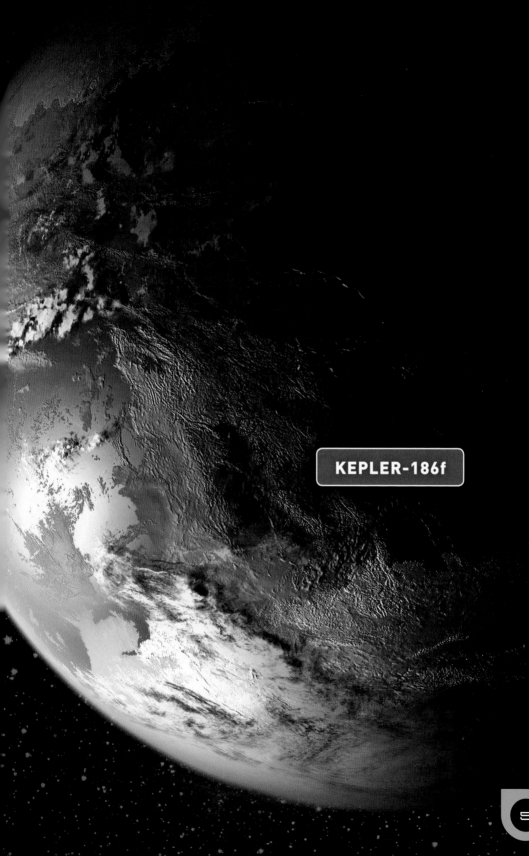

KEPLER-186f

WHAT ARE PLANETS?

Planets are objects that **orbit** a star. In Earth's solar system, the planets orbit the Sun.

Some planets are made of rocks. They may contain metals and **minerals**. Other planets are made of gases. **Gravity** keeps the rocks and gases together in a round shape.

SUN

EARTH

DWARF PLANETS

Dwarf planets are smaller than regular planets. In 2006, scientists announced that Pluto is too small to be a planet. It is now a dwarf!

EARTH

Planets can be different sizes. Some are smaller than Earth. Others are much larger. For example, Jupiter's **diameter** is more than 11 times larger than Earth's!

JUPITER

All planets must have enough **mass** to form a **sphere**. They must also be massive enough for their gravity to capture or push away nearby objects.

HOW DO PLANETS FORM?

In Earth's solar system, planets formed billions of years ago from huge clouds of gas and dust. In time, the dust clumped together to form rocky planets. These stayed close to the Sun.

Energy pushed giant gas and ice planets further away from the Sun. They formed in the cooler outer areas of the solar system.

FUN FACT

EXTRAORDINARY EXOPLANETS

Scientists have spotted many unusual exoplanets.
For example, Kepler-16b was discovered in 2011.
It is the first planet found to orbit two stars!

Over time, some planets developed an
atmosphere. Atmospheres protect planets
from energy that can damage them. They also
protect from **impacts**.

Scientists believe exoplanets formed in the
same way as the planets in Earth's solar system.
But they are still learning about other solar
systems and their planets.

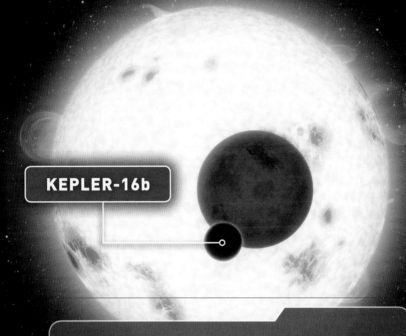

KEPLER-16b

KEPLER-16b PLANET PROFILE

Type of planet: gas giant
Size: 65,491 miles
   (105,398 kilometers) across
Stars: Kepler 16A and Kepler 16B
Orbital period: 229 days
Temperature:
   -100 to -150 degrees Fahrenheit
   (-73 to -101 degrees Celsius)

WHERE ARE PLANETS FOUND?

Planets are found in orbit around their star. In Earth's solar system, this star is the Sun. **Light-years** from Earth, exoplanets orbit stars, too!

Some planets are very close to their star. They may take less than a day to circle it. Others take more than 100 years to make one orbit!

K2-33b ORBITING ITS STAR

PLANET ORBITAL PERIODS

Mercury: 88 days
Venus: 225 days
Earth: 365 days
Mars: 1.9 years

Jupiter: 11.9 years
Saturn: 29.5 years
Uranus: 84 years
Neptune: 165 years

FUN FACT

THE SEARCH CONTINUES

So far, scientists have discovered more than 40 planets in habitable zones. They study these planets for signs of life!

Most planets are too harsh for plants and animals to survive. But each star has a habitable zone. Scientists search these zones for planets similar to Earth.

KEPLER-452b

Planets in the habitable zone are not too hot
or too cold. They may be able to support life!

WHY DO WE STUDY PLANETS?

Studying planets helps scientists understand the universe. **Telescopes** help them look at nearby planets. These devices give up-close looks at planets' **craters** and moons!

Scientists can look far away, too. Powerful instruments like the Hubble Space Telescope offer views beyond our solar system. The information they collect helps scientists make important discoveries.

HUBBLE SPACE
TELESCOPE

No life has ever been discovered
beyond Earth. But astronomers around
the world continue to look for signs.
They study planets' temperatures
and weather conditions. They also look
for water. Scientists may one day find a
planet with all the ingredients for life!

ILLUSTRATION OF TRAPPIST-1f

GLOSSARY

astronomers–people who study space

atmosphere–the gases that surround Earth and other planets

craters–deep holes in the surface of a planet or other object

diameter–the measurement of a straight line passing through the center of a circle

exoplanet–a planet outside of Earth's solar system

gravity–the force that pulls objects toward one another

habitable zone–an area around a star where the temperature may be suitable for life

impacts–events in which objects hit one another

light-years–units of length equal to the distance that light travels in one year; one light-year is 5.88 trillion miles.

mass–a measurement of how much matter an object is made up of

minerals–naturally occurring substances found in rocks, sands, and soils

orbit–to move around something in a fixed path

sphere–a ball-shaped object

telescopes–instruments used to view distant objects in outer space

TO LEARN MORE

AT THE LIBRARY

Hawksett, David. *Beyond the Asteroid Belt: Can You Explore the Outer Planets?* New York, N.Y.: PowerKids Press, 2018.

Morey, Allan. *The Hubble Space Telescope.* Minneapolis, Minn.: Bellwether Media, 2018.

Simon, Seymour. *Exoplanets.* New York, N.Y.: HarperCollins, 2018.

ON THE WEB

Learning more about planets is as easy as 1, 2, 3.

1. Go to www.factsurfer.com

2. Enter "planets" into the search box.

3. Click the "Surf" button and you will see a list of related web sites.

With factsurfer.com, finding more information is just a click away.

INDEX

The images in this book are reproduced through the courtesy of: Milissa4like, front cover, pp. 3, 4, 6, 8, 10, 12, 14, 16, 18, 20, 23 (graphic); sakkmesterke, front cover (planet); Alan Uster, front cover, pp. 2-3 (Earth/Moon); NASA/JPL-Caltech, pp. 2, 13, 14-15, 20-21; Egyptian Studio, pp. 4-5, 18-19; CHAINFOTO24, pp. 6-7 (background); Vadim Sadovski, p. 7; Triff, pp. 8-9, 16-17; Tristan3D, pp. 8 (Earth), 8-9 (Jupiter); HomeArt, pp. 10-11; Avigator Thailand, pp. 12-13 (background); NASA/Ames/JPL-Caltech, pp. 16-17 (exoplanet); Nerthuz, p. 19 (telescope inset); xtock, p. 19 (Earth inset).